Sponsored Technology Workshops 2019

AA009298

Held at the 2019 AIChE Spring Meeting and
15th Global Congress on Process Safety

New Orleans, Louisiana, USA
31 March - 4 April 2019

ISBN: 978-1-5108-8701-5

Printed from e-media with permission by:

Curran Associates, Inc.
57 Morehouse Lane
Red Hook, NY 12571

Some format issues inherent in the e-media version may also appear in this print version.

Printed by Curran Associates, Inc. (2019)

For permission requests, please contact AIChE
at the address below.

AIChE
120 Wall Street, FL 23
New York, NY 10005-4020

Phone: (800) 242-4363
Fax: (203) 775-5177

www.aiche.org

Additional copies of this publication are available from:

Curran Associates, Inc.
57 Morehouse Lane
Red Hook, NY 12571 USA
Phone: 845-758-0400
Fax: 845-758-2633
Email: curran@proceedings.com
Web: www.proceedings.com

TABLE OF CONTENTS

The Benefits of a Baseline Hazardous Area Classification Assessment for a Facility with Limited Resources

Bath Group Process Safety Management
Jim Johnston, P.E. Business Principal
Carol Ann Eccleston Sr. Process Safety Engineer
5656 South Staples, Suite 110
Corpus Christi, Texas 78411
Phone (361) 992-2284

Bath GROUP

April 2019
1

Presented at 2019 AIChE Spring Meeting & 15th GCPS

Bath Group
Process Safety Management
Booth #112

Bath GROUP Process Safety Management

April 2019
2

Presentation 3 P's

Purpose: Present the case for having a baseline Hazardous Area Classification (HAC) assessment

Process: Review slides with attendees in open Q&A format

Product: Value of baseline HAC documents in maintaining HAC compliance via a team approach

Bath GROUP Process Safety Management – *Booth #112* April 2019
3

Talking Points

- Review a Hazardous Area Classification (HAC) Case History
- Steps to resolve HAC issues
- Benefits of a baseline HAC assessment
- Takeaways

Bath GROUP Process Safety Management – *Booth #112* April 2019
4

HAC Case History

- Chemical production site in operation since 1970 and site ownership had changed several times.
- Site implemented OSHA 29 CFR 1910.119 in 1992 and recognized "Electrical classification" (i.e. HAC) as required process safety information (PSI).
- HAC responsibility was assigned to Electrical Engineering group who reevaluated HAC only on a project by project basis for design and construction input.

HAC Case History (cont.)

- Over time HAC drawings for a unit or area were only <u>found</u> by:
  1. Tracking down and asking the Electrical Engineer with greatest longevity and/or unit or area project experience for drawing numbers.
  2. Piecing together information from several projects since only HAC for project-related equipment was shown.
  3. Redoing drawing search repeatedly since some showed "Proposed" equipment for future projects which may or may not have been installed.

HAC Case History (cont.)

- Using the latest HAC drawings was challenged by:
  1. Cluttered drawings since Project-related information (e.g. instrument references, piping specs, etc.) was included.
  2. Inconsistent drawing formatting and reference notes.
  3. No associated documentation stating how the HAC assessment was done.
  4. Latest version of HAC-related Recommended and Generally Accepted Good Engineering Practices (RAGAGEP) no longer applied to some drawings.

HAC Case History (cont.)

- Then one Electrical Engineer retired and one Electrical Engineer left the company.
- This resulted in:
  1. Loss of 40 years combined experience at site.
  2. Loss of key electrical knowledge for 20 major projects at site.
  3. Confusion for new Electrical Engineers.
  4. Stress on organization since managing and maintaining HAC drawings was individual-based and not team-based.

HAC Case History Conclusions

Q: Why is it so difficult to manage and maintain HAC drawings as PSI?

A: Four (4) key reasons impede viewing HAC as PSI:

1. **HAC missing from baseline PSI documents due to:**
   - Not understanding what "Electrical classification" is
   - Thinking HAC documentation is a "once and done" exercise
   - Gaps in documentation due to Facility sale, once or multiple times

Bath GROUP Process Safety Management – *Booth #112* April 2019 9

HAC Case History Conclusions (cont.)

A: Four (4) key reasons impede viewing HAC as PSI (cont.):

2. **Perception only Electrical Engineer can determine and assign HAC (Class, Division, Material Group, and T Code/Ignition Temperature).**
3. **Lack of understanding why, how, and when to perform an HAC assessment.**
4. **Transition of personnel with OSHA process safety management (PSM) knowledge (14 elements).**

Bath GROUP Process Safety Management – *Booth #112* April 2019 10

Steps To Resolve HAC Issues

1. Understand purpose of HAC within context of PSM.
2. Create site baseline HAC documentation (HAC study and HAC compliance assessment) using "Common Sense Approach".
3. Leverage interrelationship between OSHA's 14 basic PSM elements.
4. Communicate HAC-related PSI via a team approach.

Benefits of This Approach

- Increased overall understanding of PSM.
- Standardized baseline HAC documentation easier to modify when changes or transitions occur.
- Shared ownership of HAC information rather than with a single discipline – *eliminate silos*.
- Improved communication of HAC-related PSI.
- Reduced time maintaining HAC-related PSI.

Steps To Resolve HAC Issues

1. **Understand purpose of HAC within context of PSM.**
2. Create site baseline HAC documentation (HAC study and HAC compliance assessment) using "Common Sense Approach".
3. Leverage interrelationship between OSHA's 14 basic PSM elements.
4. Communicate HAC-related PSI via a team approach.

Overview of HAC

- OSHA 29 CFR 1910.119 "Process safety management of highly hazardous chemicals" became US law in 1992:
  - Listed specific documents as required process safety information (PSI)
  - Required PSI to be updated if process changed (i.e. kept "evergreen")
  - Required PSI to be formally reevaluated every 5 years

Overview of HAC (cont.)

- Electrical Classification included on OSHA's PSI list.
- Word "electrical" in Electrical Classification is correct but misleading.
- Term focuses on how it is applied rather than how it is determined.
- Industry is moving toward preferred term "hazardous area classification" (HAC).

Bath GROUP Process Safety Management – *Booth #112* April 2019 15

Overview of HAC (cont.)

- Intent of HAC is to minimize risk of electrical energy as a potential source of ignition due to a non-catastrophic release of flammable material - *does not address static electricity.*
- HAC requirements given in NPFA 70® National Electrical Code® (NEC®) so thought to be "owned" by electrical-based disciplines (i.e. Electrical Engineer).

Bath GROUP Process Safety Management – *Booth #112* April 2019 16

Overview of HAC (cont.)

- Determined per methodologies in applicable HAC-related Recognized And Generally Accepted Good Engineering Practices (RAGAGEP).
- Results based on composition of process streams within operating unit.
- Process chemistry and operational knowledge required to perform HAC assessment.

Overview of HAC (cont.)

- Overall HAC process includes:
  1. Performing HAC assessment.
  2. Documenting HAC study results (report and drawings).
  3. Completing HAC compliance assessment.
- HAC compliance key in maintaining overall site process safety.
- HAC compliance assessment provides feedback loop for continual improvement.

Steps To Resolve HAC Issues

1. Understand purpose of HAC within context of PSM.
2. **Create site baseline HAC documentation (HAC study and HAC compliance assessment) using "Common Sense Approach".**
3. Leverage interrelationship between OSHA's 14 basic PSM elements.
4. Communicate HAC-related PSI via a team approach.

Bath GROUP Process Safety Management – *Booth #112* April 2019 19

HAC Assessment

- Results based on probability a flammable mixture will exist in air if released outside its normal containment (i.e. process equipment).
- Applicable to a manufacturing or process area.
- Requires process specific information.
- "Common Sense Approach" balances intent of RAGAGEP with best available tools to determine extent of flammable or combustible area.

Bath GROUP Process Safety Management – *Booth #112* April 2019 20

HAC Assessment (cont.)

- HAC assessment results must include four (4) components:
  - Class:
    - Class I for Vapors /Gases/Liquids
    - Class II for Combustible Dusts
    - Class III for Ignitable Fibers and Flyings
  - Division (1 or 2)
  - Material Group (specific to Class)
  - Maximum Surface Temperature (NEC® T Code) (Class I) or Ignition Temperature (Class II)

Bath GROUP Process Safety Management – *Booth #112* April 2019 21

HAC Assessment (cont.)

- Methodologies require knowledge of the process chemistry.
- PSI required to complete HAC assessment includes:
  - Material Balance (compositions, flow rates, P, T)
  - Flash point, autoignition temperature, vapor density
  - Minimum ignition energy, dust cloud ignition temperature, electrical resistivity, dust layer thickness
  - Unit operations specific to process equipment
  - P&ID's, PFD's, plan and elevation drawings

Bath GROUP Process Safety Management – *Booth #112* April 2019 22

HAC Compliance Assessment

- Required by OSHA and the NEC®.
- Covers installed electrical equipment and electrical installation practices.
- Gap analysis between What's there vs. Required.
- Results can identify:
  - Additional cost for compliance prior to site changes
  - Systemic electrical equipment implementation issues
  - Problems with document control process

Baseline HAC Documentation

- Baseline HAC drawings show classification visually.
- HAC drawings are primary at point-of-use and HAC report is secondary (background information).
- HAC assessment report gives bases for the assigned classification (i.e. Why and How).
- "Common Sense" approach puts main emphasis on making HAC drawings clear so all individuals can obtain HAC-related information quickly.

Steps To Resolve HAC Issues

1. Understand purpose of HAC within context of PSM.
2. Create site baseline HAC documentation (HAC study and HAC compliance assessment) using "Common Sense Approach".
3. **Leverage interrelationship between OSHA's 14 basic PSM elements.**
4. Communicate HAC-related PSI via a team approach.

Bath GROUP Process Safety Management – *Booth #112* April 2019
25

OSHA's 14 Basic PSM Elements

Bath GROUP Process Safety Management – *Booth #112* April 2019
26

Leverage OSHA's PSM Requirements

- Understand PSI requirements for each element and how each element relates to another.
- Issue PSI updates in a timely manner via document management system - *critical.*
- Use same PSI source to eliminate duplication of effort.
- Drive an efficient PSM communication process via true team approach – *clear roles & responsibilities and expectations.*

Steps To Resolve HAC Issues

1. Understand purpose of HAC within context of PSM.
2. Create site baseline HAC documentation (HAC study and HAC compliance assessment) using "Common Sense Approach".
3. Leverage interrelationship between OSHA's 14 basic PSM elements.
4. **Communicate HAC-related PSI via a team approach.**

HAC-Related PSI

- NEC® covers HAC requirements:
  - How to determine (study inputs & RAGAGEP)
  - What to document (study basis & drawings)
  - Who should use results (specific roles)
- NEC® states HAC "Documentation shall be available to those authorized to design, install, inspect, maintain, or operate electrical equipment at the location".

HAC-Related PSI (cont.)

- HAC results required to safely manage electrical equipment and electrical systems in a production unit.
- HAC results to be utilized by support Team members:
  - Operations
  - Process Engineering
  - Project Engineering
  - Maintenance
  - Mechanical Integrity
  - Construction

HAC-Related PSI (cont.)

- HAC-related PSI utilized by PSM elements such as:
  - <u>Plot plan</u> for Emergency Response, Compliance Audit, Training
  - <u>Material Balance</u> for Process Hazard Analysis, MOC, Training, Trade Secrets
  - <u>P&ID's</u> for Operating Procedure, PSSR, Incident Investigation, Mechanical Integrity
  - <u>Operating conditions</u> for Hot Work, Operating Procedures, Contractors

HAC Information Utilization Process

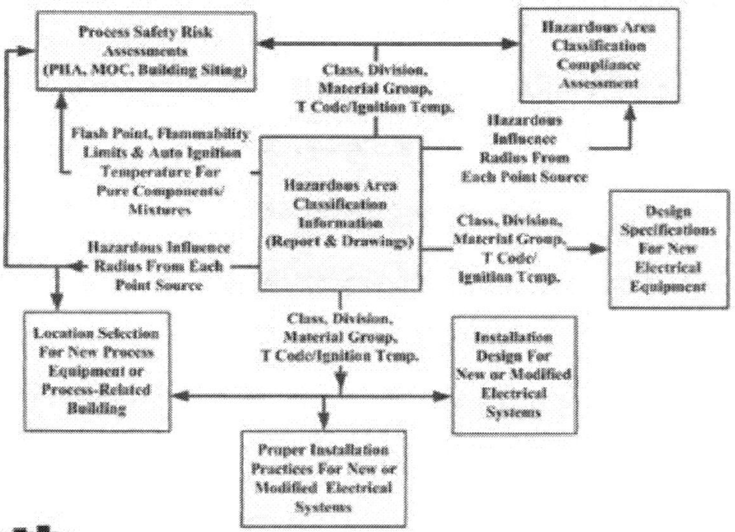

Takeaways

- Process changes in facilities are commonplace.
- HAC-related PSI is routinely impacted as a result.
- Baseline HAC documentation (report and drawings) uses latest versions of PSI.
- Baseline HAC report allows efficient review to identify changes as part of MOC and PHA studies.
- Baseline HAC drawings allow project-related drawings to be created without revising "record" version.

Bath GROUP Process Safety Management – *Booth #112* April 2019
33

Takeaways (cont.)

- NEC® requires HAC documentation (report and drawings) be available to those authorized to design, install, inspect, maintain, or operate electrical equipment.
- Implement Team-based approach for shared ownership of HAC-related PSI to leverage:
  - Understanding
  - Knowledge
  - Utilization

Bath GROUP Process Safety Management – *Booth #112* April 2019
34

Takeaways (cont.)

- This team approach is critical in facilities with limited resources to:
  - Managing an effective PSM process
  - Maintaining HAC compliance
- Bath PSM website contains instructional videos covering many HAC-related topics at **https://www.bathpsm.com/instructional-videos**.

Bath GROUP Process Safety Management – *Booth #112* April 2019 35

Remember, a Baseline HAC Assessment Is

Common Sense vs.

Non-sense

Bath GROUP Process Safety Management – *Booth #112* April 2019 36

579048 Batch Process Modeling - Accelerating Innovation to Remain Competitive

The batch processing industries are taking advantage of simulation to accelerate innovation and optimize production of some of their most challenging products. Explore how using Aspen Plus, you accomplish quick and easy modeling of your most challenging batch processes. The Path for digital transformation.

579056 Industry 4.0: Digitalization in Chemical Industry

Digital solutions are key to driving innovation and new product introduction in the competitive specialty chemicals market. Digital transformation can be implemented in many business processes and have several maturity levels eliminating the need to jump into your digitalization journey thinking the highest level of maturity is the only path to value. Explore how to bring products to market faster, improve product throughput as well as quality, and wring insight from plant data. Uncover the value of digitalization from innovation through operations.

AUTHOR INDEX